SPOTLIGHT ON PHYSICAL SCIENCE

WAVES

EXAMINING INTERACTIONS

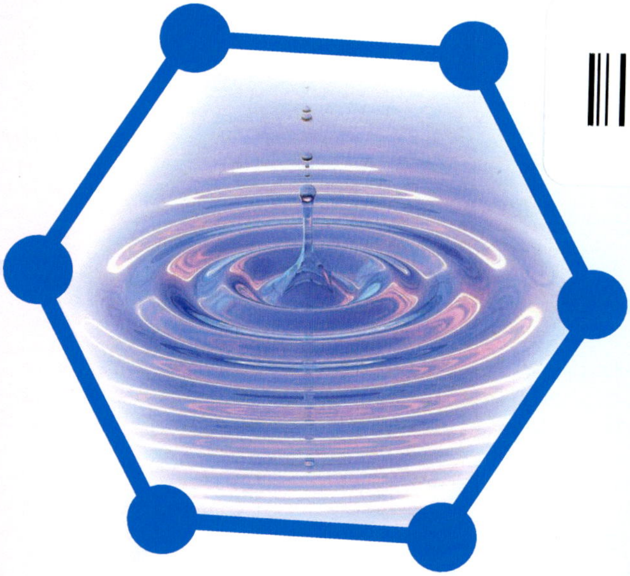

AMANDA VINK

PowerKiDS press.
NEW YORK

Published in 2020 by The Rosen Publishing Group, Inc.
29 East 21st Street, New York, NY 10010

Editor: Jennifer Lombardo
Book Design: Michael Flynn

Photo Credits: Cover, p. 1 Corona Borealis Studio/Shutterstock.com; (series molecular background) pro500/Shutterstock.com; p. 5 (ocean) Kelly Headrick/Shutterstock.com; p. 5 (puddle) Joshua Small-Photographer/Shutterstock.com; p. 6 Daniel Krason/Shutterstock.com; p. 7 Vladimir Nenezic/Shutterstock.com; p. 9 https://commons.wikimedia.org/wiki/File:Lyre_player_Met_06.1021.188.jpg; p. 11 DEA/G. NIMATALLAH/De Agostini/Getty Images; p. 13 © iStockphoto.com/Ben Gingell; p. 14 Trofimov Denis/Shutterstock.com; p. 15 Vladimir Gjorgiev/Shutterstock.com; p. 17 (people) Maridav/Shutterstock.com; pp. 17 (diagram), 26 Fouad A. Saad/Shutterstock.com; p. 19 EML/Shutterstock.com; p. 19 Kicky_princess/Shutterstock.com; p. 20 Stefano Bianchetti/Shutterstock.com; p. 21 https://en.wikipedia.org/wiki/Albert_Einstein#/media/File:Albert_Einstein_Head.jpg; p. 23 Fancy Tapis/Shutterstock.com; p. 25 Kostenko Maxim/Shutterstock.com; p. 27 sw_photo/Shutterstock.com; p. 29 https://commons.wikimedia.org/wiki/File:Christian_Doppler_(1803-1853),_Nr._111,_bust_(marble)_in_the_Arkadenhof_of_the_University_of_Vienna-2934.jpg; p. 30 vchal/Shutterstock.com.

Library of Congress Cataloging-in-Publication Data

Names: Vink, Amanda, author.
Title: Waves : examining interactions / Amanda Vink.
Description: New York : PowerKids Press, [2020] | Series: Spotlight on physical science | Includes bibliographical references and index. | Summary: "The ringing sound of an electric guitar, the ripples after you've thrown a rock in the water, even the way a microwave oven cooks your food-these are all examples of what waves can do! A wave is one way that energy travels through space and matter. This book breaks down complex physical science topics so young readers can understand them more easily. Real-life examples of how people use waves for science, art, and more help readers see how waves are all around us"-- Provided by publisher.
Identifiers: LCCN 2019029726 | ISBN 9781725313446 (paperback) | ISBN 9781725313477 (library binding) | ISBN 9781725313453 | ISBN 9781725313484 (ebook) | ISBN 9781725313460
Subjects: LCSH: Waves--Juvenile literature.
Classification: LCC QC157 .V56 2020 | DDC 531/.113--dc23
LC record available at https://lccn.loc.gov/2019029726

Manufactured in the United States of America

CPSIA Compliance Information: Batch #CWPK20. For further information contact Rosen Publishing, New York, New York at 1-800-237-9932.

CONTENTS

CHAPTER ONE

WAVES MOVE ENERGY!

Have you ever looked closely at water? Maybe you've stood next to an ocean or jumped in a puddle. If so, you've probably seen ripples. You may have even seen a wave in the water as it curls and makes its way towards the shore.

When water moves, it's easy to see waves at work. In addition to waves in the water, there are many other types of waves in our physical world. Light waves and sound waves, for example, help control the way humans experience the world. Understanding waves better helps us understand physical science, engineering, and other things about our world.

What is a wave, exactly? Waves are **disturbances** that travel from one location to another, sometimes through a **medium**. A wave is the shift of energy. After the disturbance has passed, the medium returns to the state it was in before.

There are many waves in our world. Water waves are some of the easiest to spot!

CHAPTER TWO

WHAT IS ENERGY?

Energy is the ability to do work, or the action of moving something against a force. In **physics**, energy causes movement. The total amount of energy in a closed system, such as the universe, can never be changed. Energy can't be created nor destroyed. It can only be moved or changed from one thing to another.

A metal spring has more potential energy when it's stretched or compressed.

There are many different types of energy, but they fall into two main groups. Potential energy is stored energy, or energy not in use yet. A metal spring at rest has less potential energy than a spring that has been compressed or stretched out. Kinetic energy, on the other hand, is the energy of moving objects. A spring that's just been released has more kinetic energy than one that is still stretched or compressed. Within these two groups, there are different kinds of energy, including thermal, chemical, nuclear, and electrical energy.

CHAPTER THREE

STARTING WITH SOUND

Our current understanding of waves began with the study of **acoustics** and music. Pythagoras, who lived in Greece around 500 BC, knew that vibrating strings created sound, and he worked to understand the notes and scales used in Greek music. The first thing he found out was simple: two strings of the same length, **tension**, and thickness will have the same sound, or pitch. Strings that are different from each other have different pitches. Pythagoras called the relationship between two different notes an interval.

In the fourth century BC, another Greek man, Aristotle, realized that a sound wave travels through air by motion. A Roman man named Marcus Vitruvius Pollio studied the Greeks' research on sound waves and wrote about how their ideas could be used to build theaters that would carry sound well. Italian Galileo Galilei also published a clear statement about vibration and sound in the 17th century.

The man in the image on this piece of pottery is holding a lyre, a stringed instrument often used in ancient Greece.

CHAPTER FOUR

NEWTON'S LAWS OF MOTION

Waves transfer, or move, energy, but not mass, which means the medium a wave is moving through doesn't travel. It's a little like doing "the wave" at a sports match. Even though the wave is traveling around the arena when people stand, the people themselves are not moving around the arena.

During the 1600s, English mathematician Isaac Newton explained his theories about the physical laws of the world, which included his three laws of motion.

Newton's first law says every object will remain at rest or moving in a straight line unless a force interrupts it. In other words, without any forces such as gravity acting upon them, objects tend to keep doing what they're doing. This is called inertia. Newton's theory was **radical** at the time. For 2,000 years before that time, most people believed objects tended to come to a resting position on their own.

Sir Isaac Newton is one of the most famous individuals of the Scientific Revolution, which marked the beginning of modern-day scientific reasoning.

Newton's second law of motion states that when a force acts on an object to make it move, how quickly the object speeds up is the result of how much force is used. However, objects with more mass move more slowly than objects with less mass. If you used the same amount of energy to throw a tennis ball and then a basketball, which has a greater mass, the tennis ball would move more quickly than the basketball.

The third law of motion says that for every action in nature, there is an equal and opposite reaction. For example, if air rushes out of a balloon, the balloon flies up in the opposite direction unless something is holding it in place.

These laws are important because they show people how natural occurrences in the world follow mathematical principles. Waves also follow these principles. For example, until it hits a shoreline or something else, a ocean wave will continue to move.

A basketball has a greater mass than a tennis ball. Mass is constant, unlike weight, so an object's mass is the same on the moon as it is on Earth.

CHAPTER FIVE

MECHANICAL OR ELECTROMAGNETIC?

Waves are all around us. Humans can see some waves but not others. All waves are either mechanical waves, which are waves that require a medium, or electromagnetic waves, which are waves that can travel through a vacuum, or empty space.

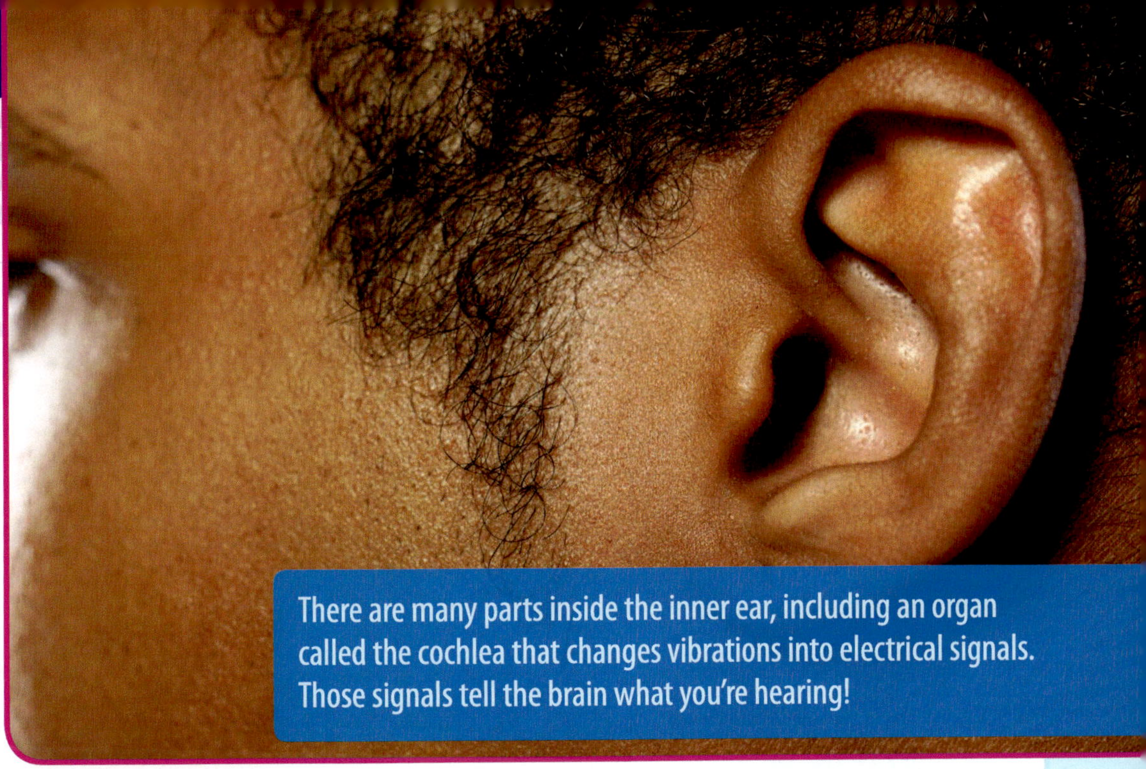

There are many parts inside the inner ear, including an organ called the cochlea that changes vibrations into electrical signals. Those signals tell the brain what you're hearing!

Mechanical waves, such as sound waves, travel when molecules in a medium bump up against one another. Sound travels through air, water, and solids, but it can't travel through a vacuum. When the movement of the molecules hits parts in our ears, we can hear the world around us.

Electromagnetic waves, such as light waves, travel through the electrical and magnetic fields that charged particles—very tiny pieces of matter—**generate**. Rays from the sun, for example, travel through empty space and finally reach Earth. It takes about 8 minutes and 20 seconds for sunlight to travel here from the sun.

CHAPTER SIX

THREE MECHANICAL WAVES

Mechanical waves are grouped by the direction the disturbance travels. With transverse waves, the particles move **perpendicular** to the direction of the wave. A good example of a transverse wave is a rope that is moved in a snakelike pattern.

With longitudinal waves, the disturbance moves in the same direction as the wave. A good example of this is a spring, which **distributes** energy in the same direction in which it moves. Sound waves are also longitudinal.

The last kind of mechanical wave is called a surface wave. Particles in a surface wave vibrate both up and down and back and forth, so they move in a circle. A wave traveling on the surface of the ocean is a good example of a surface wave. The individual particles move in a circle.

Transverse waves move perpendicular to the source, or the thing that is making the waves. Longitudinal waves move in the same direction as the source.

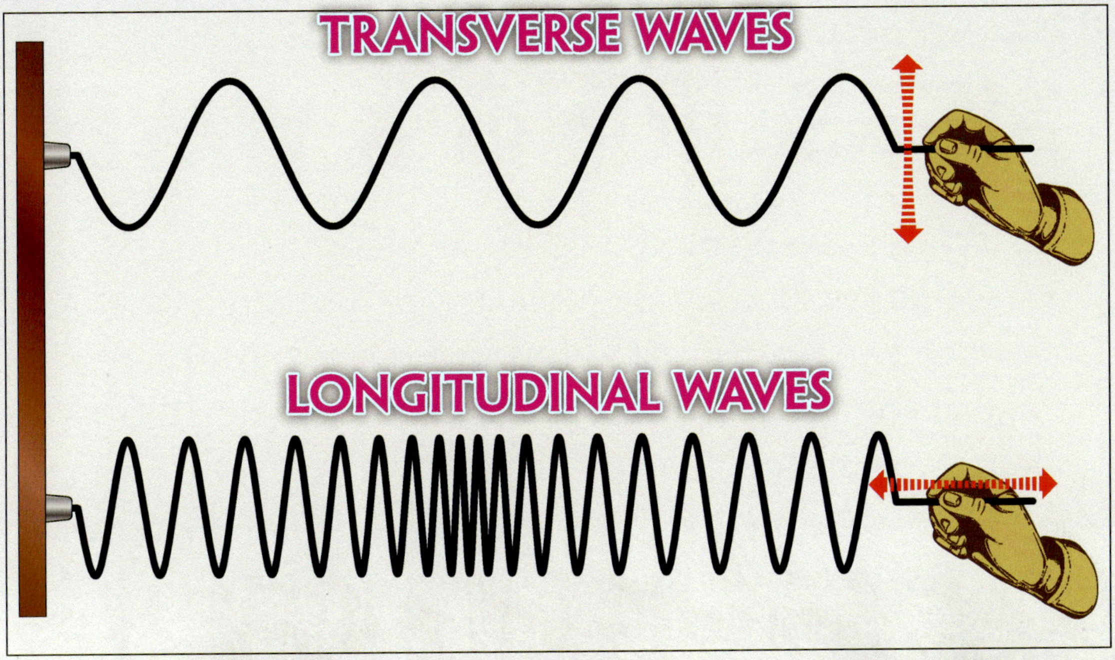
TRANSVERSE WAVES
LONGITUDINAL WAVES

CHAPTER SEVEN

MEASURING A WAVE

Waves can be measured in wavelengths. The wavelengths of longitudinal waves are measured between two points that are either both in the same stage of compression (being pushed together) or the same stage of stretching out.

The wavelengths of transverse waves are measured from crest to crest or trough to trough. The crest is the highest point of the wave and the trough is the lowest point. A wave's amplitude is the distance from the middle of the wave to the top of a crest or bottom of a trough.

Waves can travel great distances. Light waves, for example, travel through space all the way from the sun and other stars. A single water wave can travel across an entire ocean.

The medium through which a wave travels often affects how the wave travels. For example, sound waves travel through air, and they travel outward in all directions.

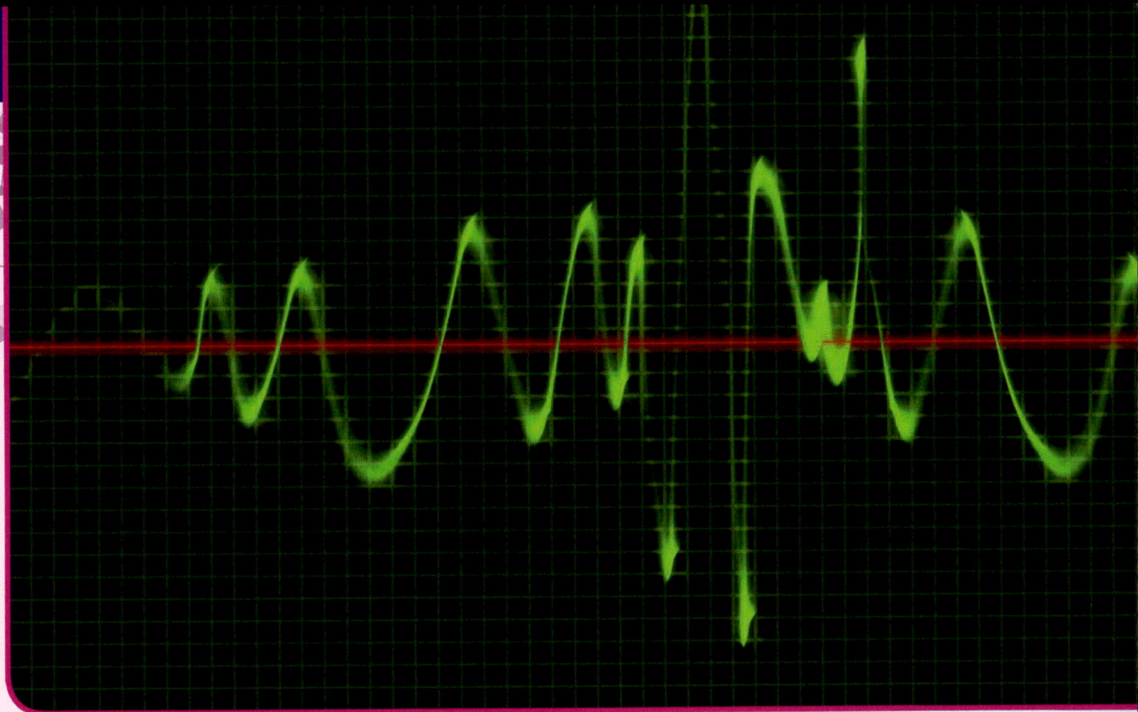

PARTS OF A WAVE

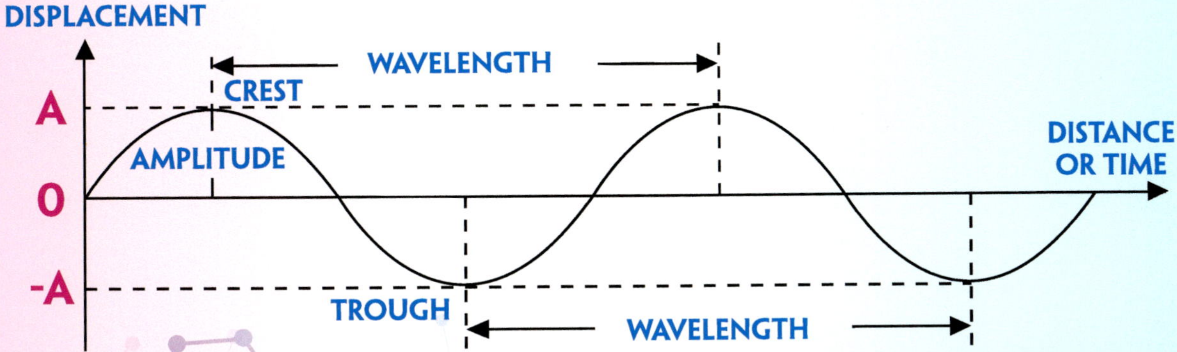

This diagram shows the parts of a wave.

CHAPTER EIGHT

ELECTROMAGNETIC WAVE THEORIES

For much of the 19th century, physicists weren't sure how light works. Many scientists believed light was a wave, and many others believed light was a particle.

In the 1860s—long before Einstein—James Clerk Maxwell, a Scottish man who studied physics, came up with a theory that light was an example of

JAMES CLERK MAXWELL

In 1905, physicist Albert Einstein proposed a theory that light acted as both a wave and a particle. He won the Nobel Prize in physics in 1921.

electromagnetic **radiation**. In 1886, physicist Heinrich Hertz became the first person to prove that waves carry electromagnetic energy between two places. Hertz was interested in Maxwell's theory. He began creating electric sparks using an induction coil, a device that creates **voltage**. After many experiments, Hertz created and detected radio waves. Soon, people invented radios and other devices to use these waves. By 1901, radio waves crossed the Atlantic Ocean from Britain to Canada.

CHAPTER NINE

TYPES OF ELECTROMAGNETIC WAVES

The waves on the electromagnetic **spectrum** are organized according to their frequency, or the number of waves that pass a fixed point in a certain amount of time. They can also be organized according to their wavelength. The greater the energy of a wave is, the larger its frequency is and the smaller its wavelength is.

Visible light, or light humans have the ability to see, is located in the middle of the spectrum. Gamma rays, X-rays, and ultraviolet (UV) light have higher frequencies than visible light. Infrared rays, microwaves, and radio waves have lower frequencies than visible light. Radio waves have the longest wavelengths. Their wavelengths can be several miles long! Most electromagnetic radiation isn't visible to humans, but humans have been able to figure out ways to use these waves anyway. For example, scientists use UV light to study crime scenes.

ELECTROMAGNETIC SPECTRUM

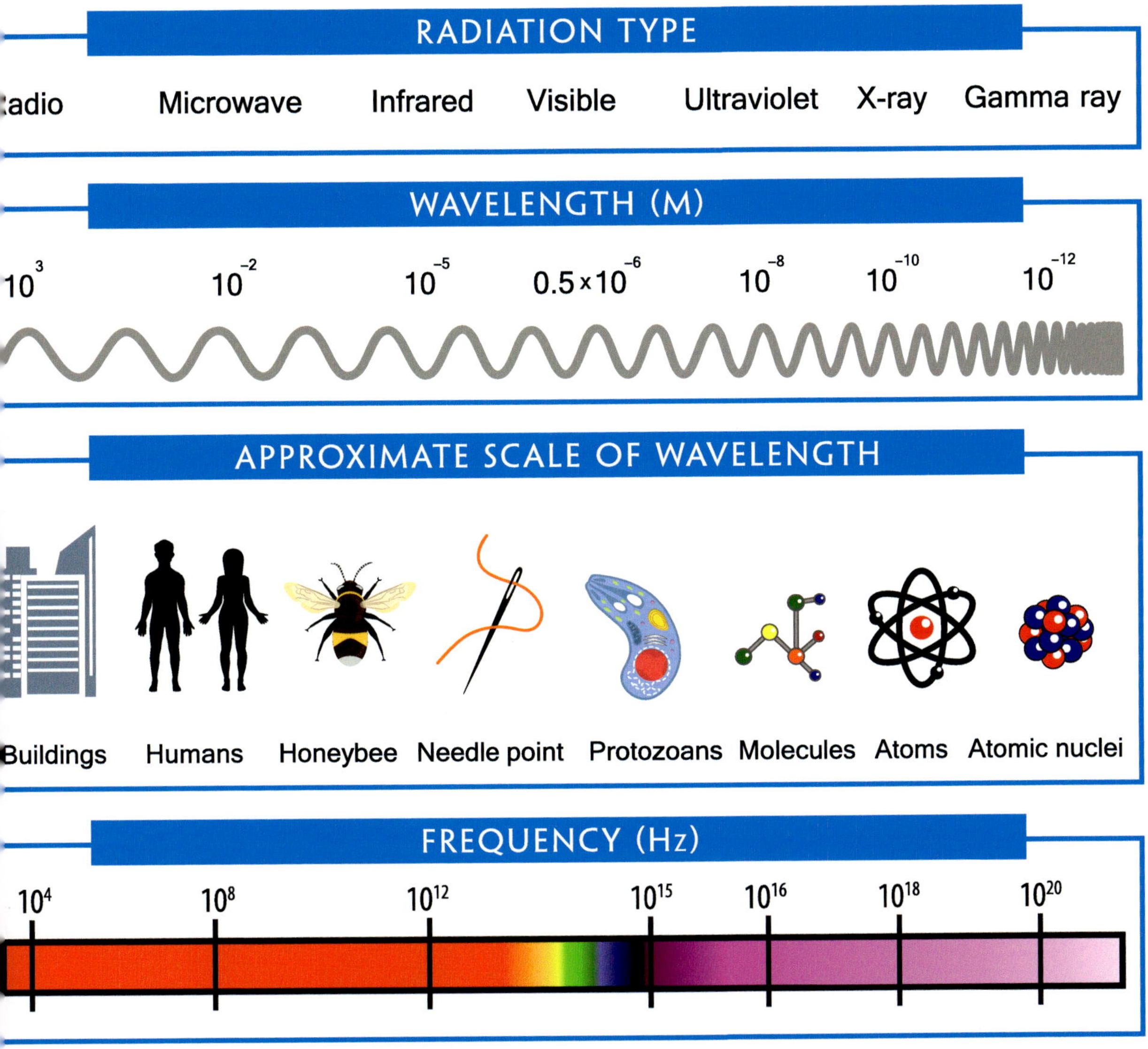

Waves in the electromagnetic spectrum are sorted by wavelength and frequency.

CHAPTER TEN

WAVE BEHAVIOR

Waves can go through reflection, refraction, diffraction, and interference. If you've looked into a mirror and seen yourself, you've experienced reflection! This happens when a wave hits a boundary and suddenly changes directions. A reflected wave will bounce off an object at the same angle as it hits it.

Refraction is when a wave bends as it enters a different medium in which it moves faster or slower. For example, sunlight may refract into a rainbow when it hits water in the air during a rain shower. Each color of the light that makes up sunlight has a different wavelength, and all move at different speeds. They all bend at a slightly different angle.

Diffraction happens when a wave runs into either an object or a small opening in its path. Waves bend around the object or spread out as they pass through the opening.

Sir Isaac Newton was the first to purposefully split white light into a rainbow using a glass called a prism (shown here) so he could study it.

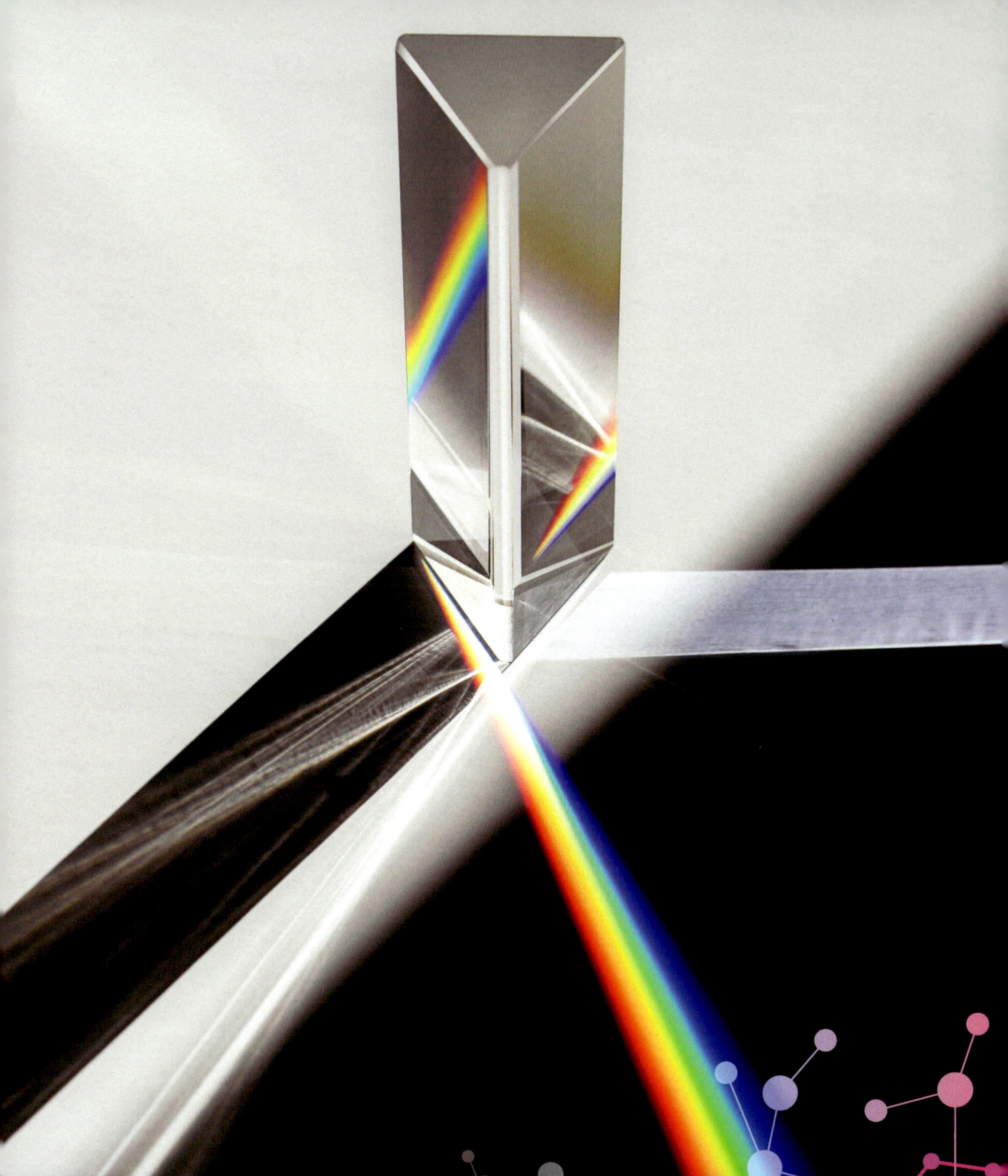

CHAPTER ELEVEN

WHEN WAVES COLLIDE

What happens when two or more waves traveling through the same medium run into one another? The interaction of waves with other waves is called wave interference.

WAVE INTERFERENCE

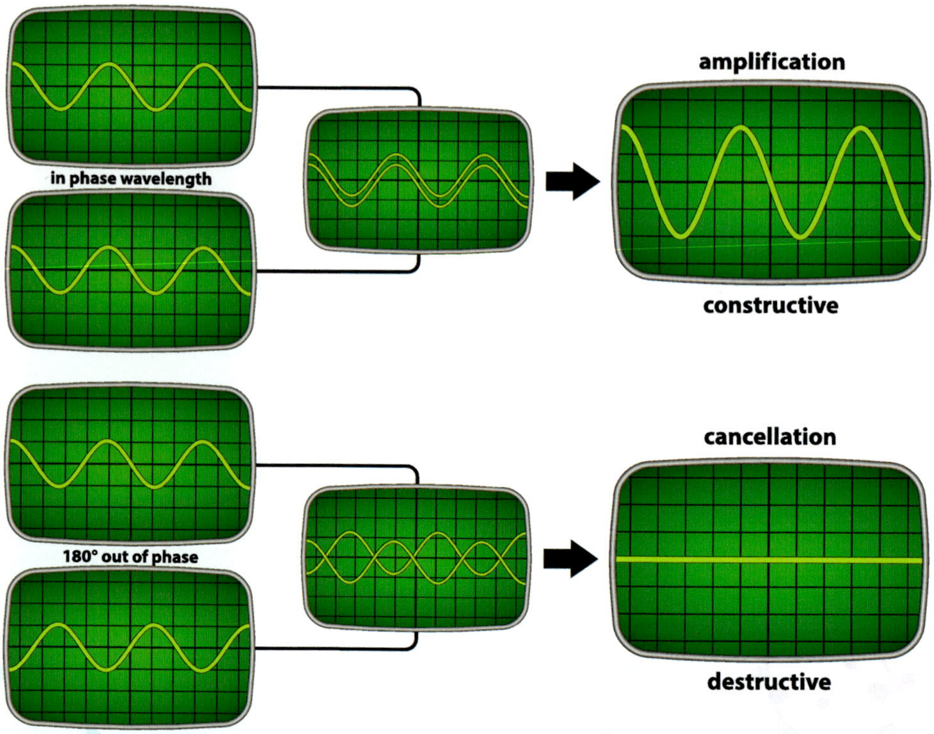

Depending on how waves hit one another, the waves can get bigger or they can cancel each other out.

When two or more traveling waves run into one another, it affects their amplitude. Interference can be called constructive or destructive. Constructive interference is when the crests of both waves overlap one another. This causes the new wave to have a greater amplitude. Destructive interference is when one wave's crest hits the other wave's trough. In this case, the crest and trough cancel each other out, and the new wave has less amplitude.

When waves hit one another, an interference pattern can form. This can be helpful. For example, surfers use interference patterns in water waves to guess which waves will be the highest. Surf's up!

CHAPTER TWELVE

THE DOPPLER EFFECT

Have you ever noticed that the sound of a fire truck's siren is different depending on where you are in relation to the moving truck? As it's approaching you, the siren sounds very high pitched. However, after it passes you and is moving away, the sound changes and becomes lower in pitch. This is because as the fire truck moves, it's catching up to its own sound, and the waves bunch up. The sound a person hears as the truck approaches them is higher pitched than the actual sound coming from the truck.

This is called the Doppler effect, and it shows how motion changes waves. Austrian scientist Christian Doppler developed this theory in 1842. Doppler also said that the color of a star changes according to the star's velocity, or speed in a given direction.

This bust of Christian Doppler by Georg Leisek is located at the University of Vienna in Austria.

CHAPTER THIRTEEN

WAVES IN THE WORLD

Knowing how waves work not only helps scientists understand events in our world, it also allows people to create many wonderful things. For example, scientists have used the measurements of **seismic** waves to understand the structure of the planet. Using a seismograph, or a machine that records the shaking of the earth, scientists can measure the mechanical waves that cause earthquakes. Knowing this information helps keep people safe by helping them figure out how to build structures that can withstand earthquakes. It also gives them more information about our planet.

Waves move energy all around us. Because of waves, you can have a conversation with a friend or enjoy the sunshine. What waves have you experienced today? Can you figure out what kind of waves they were?

GLOSSARY

acoustics (uh-KOOS-tiks) The science of sound; also, the qualities of a place that make it easy or hard for people to hear.

distribute (dih-STRIH-byoot) To share or spread out.

disturbance (dih-STUHR-buhns) A change in the position or order of something.

generate (JEH-nuh-rayt) To produce something or cause something to be produced.

medium (MEE-dee-uhm) The thing by which or through which something is done.

perpendicular (puhr-puhn-DIH-kyuh-luhr) Being at right angles.

physics (FIH-ziks) The study of matter, energy, force, and motion, and the relationship among them.

radiation (ray-dee-AY-shuhn) Waves of energy.

radical (RAH-dih-kuhl) Very new and different, not ordinary.

seismic (SYZ-mik) Having to do with an earthquake.

spectrum (SPEHK-truhm) An entire range of something, such as waves.

tension (TEHN-shuhn) How stretched something is.

voltage (VOHL-tij) The force of an electrical current.

INDEX

PRIMARY SOURCE LIST

Page 9
Terracotta kylix (drinking cup). Attributed to the Dokimasia Painter. c. 480 BC. Department of Greek and Roman Art, Metropolitan Museum of Art, New York City, New York.

Page 11
Portrait of Isaac Newton (1642-1727). Copy of painting by Godfrey Kneller. Barrington Bramley. 1992. Institute for Mathematical Sciences, University of Cambridge.

Page 29
Bust of Christian Doppler. Georg Leisek. Marble. 1901. Arkadenhof of the University of Vienna.

WEBSITES

Due to the changing nature of Internet links, PowerKids Press has developed an online list of websites related to the subject of this book. This site is updated regularly. Please use this link to access the list: www.powerkidslinks.com/SOPS/waves